2023 Annular Solar Eclipse:

Ring of Fire-Color Edition

Unique Fred

Table of Contents

Introduction

An annular eclipse, a rare and captivating celestial phenomenon, paints the sky with wonder and mystery. During this spectacle, the Moon gracefully drifts in front of the Sun, allowing a compelling ring of sunlight to encircle the Moon's dark disk.

<u>Ring of Fire</u>

This mesmerizing contrast creates a breathtaking sight, often called a "ring of fire" due to its fiery appearance. The annular eclipse of 2023 promises to be one of the most remarkable celestial events of the decade. It marks the return of an annular eclipse to the continental United States, an event last witnessed in 1994 and the first visible from North America since 2012.

Notably, it will also be the longest annular eclipse of the 21st century, with its awe-inspiring display extending at its maximum point for more than six minutes. This annular eclipse will follow a narrow path that stretches across the United States, from Oregon to Texas, before reaching parts of Mexico, Central America, Colombia, and Brazil. For an in-depth look at its journey, refer to the annular eclipse path map.

<u>Annular Eclipse Path with Best Viewing Spots</u>

However, one must approach it cautiously and prepare to embrace this celestial spectacle fully. Unlike a total eclipse, an annular eclipse doesn't obscure the Sun entirely; gazing at it without proper eye protection is hazardous. Additionally, capturing the intricate beauty of the ring of fire requires a suitable camera and lens. This book is your guide to safely observing and photographing the annular eclipse, offering invaluable tips and tricks to ensure you make the most of this unique opportunity.

Yet an annular eclipse is more than just a visual marvel; it's a scientific marvel that unveils profound insights into the Sun, the Moon, and the Earth's interactions. In this book, you'll delve into the science behind eclipses, understanding how these celestial alignments occur, their predictability, and their profound impact on our planet and beyond. As you turn the pages, you'll explore the future of eclipses in an era marked by space exploration and newfound discoveries.

This book is a comprehensive resource for anyone curious about the 2023 annular eclipse or eclipses in general. Whether you are a novice or an expert, a casual observer or an avid enthusiast, this guide equips you with all the knowledge you need to embark on a journey of wonder and understanding. Together, let's witness one of nature's most spectacular shows: the glorious Ring of fire.

Chapter 1

Intrigue

Difference

An annular eclipse and a total eclipse belong to the extraordinary family of solar eclipses, those delightful moments when the moon slips between our Earth and the radiant sun, casting a celestial shadow on our planet. However, these two mesmerizing events bring distinct characteristics owing to the ever-dancing cosmic bodies.

In the realm of annular eclipses, the moon stands near the farthest point in its elliptical orbit around our Earth, an aspect known as **apogee**. When this occurs, the moon's apparent size appears smaller than the sun's. As a result, even when these celestial spheres align with precision, the moon falls short of completely concealing the sun. The consequence is a captivating and peculiar phenomenon: a delicate, fiery ring of sunlight encircles the moon's dark, mysterious disk. This enchanting sight has earned annular eclipses the moniker "ring of fire," owing to their blazing appearance.

<u>Ring of Fire</u>

In contrast, the stage for a total eclipse is set when the moon draws near the closest point in its orbit around our Earth, a moment referred to as **perigee**. At perigee, the moon's apparent size either matches or surpasses the sun's, effectively allowing it to cloak the sun in darkness for a brief period.

During this celestial embrace, the sun's outer atmosphere, the corona, emerges as a faint, ethereal halo, delicately adorning the moon's silhouette. This spectacle, aptly named "totality," reveals the sun's hidden features and bathes the world in awe-inspiring darkness.

A Total Solar Eclipse with the Sun's Corona Visible as a Front Halo Around the Sun

A visually illuminating demonstration of the distinction between an annular and total eclipse can be observed through this.

<u>Distinction Between an Annular and Total Eclipse</u>

Both annular and total eclipses, exquisite celestial ballets, demand the perfect celestial alignment of the sun, the moon, and our Earth. Their occurrence is also influenced by the moon's constantly shifting position and distance in its orbit. In the near future, we anticipate the arrival of the next annular eclipse on October 14, 2023, painting the skies over select regions of North America, Central America, and South America. Subsequently, the stage will be set for the next total eclipse on April 8, 2024, gracing the skies over parts of North America, Central America, and South America.

What Sets it Apart

The 2023 annular eclipse is a celestial event of extraordinary significance, and its uniqueness extends beyond mere astronomical curiosity. Here, we delve into the myriad reasons that make this eclipse truly special:

★ This annular eclipse is poised to etch its historical mark. It will be a once-in-a-generation event for those residing in the continental United States. Not since 1994 has an annular eclipse graced this region. Additionally, North America will witness the spectacle for the first time in a decade.

★ **Record-Breaking Duration:** At the heart of the 2023 annular eclipse lies its claim to fame as the longest annular eclipse of the 21st century. The moon's path across the sun's radiant face will extend for over six minutes at its zenith, offering a prolonged celestial display.

★ **Scientific Expedition:** Beyond its aesthetic grandeur, the 2023 annular eclipse is a remarkable scientific inquiry and education opportunity. Esteemed organizations such as NASA, alongside various research bodies, are poised to conduct experiments and observations during this celestial spectacle.

★ **Spectacular Visual Feast:** As the moon completes its dance with the sun, millions of people across regions touched by the moon's shadow will witness the spellbinding "ring of fire" gracing the sky. Those residing outside the path of annularity will still enjoy a partial eclipse, creating a celestial panorama.

The eclipse's visual story unfolds along a narrow path, meandering from Oregon to Texas in the United States, eventually venturing across parts of Mexico, Central America, Colombia, and Brazil. For an intricate depiction of its journey, refer to this map.

<u>Annular Eclipse path Map Across America</u>

Across the breadth of the Americas, from the northern realms of Alaska to the southern stretches of Argentina, the eclipse offers a partial spectacle, as the moon partially obscures the sun's brilliance.

Partial Eclipse Across America

Timings of eclipse commencement and conclusion vary based on one's geographic location. The partial eclipse will first grace the skies at 15:03:50 UTC, with the final glimpse concluding at 20:55:16 UTC[5]. To align these moments with your local time zone, consult the user-friendly.

Safety

The art of observing and photographing an annular eclipse is a blend of precision, care, and the right equipment. Unlike a total eclipse, where the Sun is entirely veiled, an annular eclipse leaves a thin, blazing ring around the Sun's dark disk. To ensure a safe and captivating experience, consider these detailed guidelines for both observation and photography:

Observation:

1. **Protect Your Vision:** Shield your eyes from the Sun's intense rays by donning certified eclipse glasses or employing a handheld solar viewer. These specialized solar filters safeguard your vision by blocking most of the Sun's harmful rays,

allowing you to witness the eclipse without risking eye damage.

Be forewarned: regular sunglasses, eyeglasses, or improvised filters can have severe consequences, including permanent eye damage or blindness.

Photography:

2. **Equip Your Camera:** When venturing into eclipse photography, you'll need a camera equipped with a lens capable of precisely capturing the Sun's disk. A stable platform like a tripod is essential to prevent image blurring.

3. **Solar Filter for Your Camera:** A solar filter for your camera lens is indispensable to capture the ring of fire in all its glory. These filters function similarly to eclipse glasses and solar viewers, safeguarding your camera sensor.

You can acquire ready-made solar filters tailored to your lens size or craft your own using aluminized Mylar film or metal-coated glass. It is vital to avoid using neutral density or polarizing filters, as they can damage your camera sensor.

4. **Adjust Camera Settings:** A close-up of the ring of fire requires meticulous camera settings. Utilize manual mode to control your aperture (opt for a small value like f/16 or f/22), shutter speed (opt for a fast value like 1/1000s or 1/2000s), and ISO (select a low value like 100 or 200).

To focus precisely on the Sun, use the live view mode or a magnifying loupe. Capture multiple shots at varying exposures, which can later be merged in post-processing to craft a high dynamic range (HDR) image.

5. **Plan Your Sequence:** Meticulous planning is vital for a comprehensive visual record of the eclipse's phases. Utilize apps such as PhotoPills to ascertain precise timings and locations for eclipse phases in your vicinity.

This tool also helps you determine the optimal camera direction and angle. With PhotoPills' 3D augmented reality feature, you can simulate different scenarios and plan your photographic concept meticulously.

Employ a timer or an intervalometer to capture images regularly during the eclipse. These shots can then be combined in post-processing to produce a composite image.

6. **Smartphone Photography:** For those capturing the eclipse with a smartphone,

secure a clip-on solar filter for your phone's camera lens or a handheld solar viewer to position in front of your camera lens.

Ensure a stable surface or tripod to steady your device. Use either the native camera app or a third-party application that permits manual control of exposure and focus. Employ the zoom feature or a clip-on telephoto lens for a closer view of the Sun.

By adhering to these detailed guidelines, you can safely observe the annular eclipse and **document** its **mesmerizing** beauty through the lens of your camera or smartphone.

Chapter 2

The Science of Eclipses

The Science of Eclipses invites us to embark on a captivating journey into the realm of these celestial wonders. Among nature's most captivating and mesmerizing phenomena, Eclipses unfold when a perfectly orchestrated cosmic dance plays out between the Sun, the Moon, and our very own Earth.

This choreography culminates in creating a shadow that elegantly sweeps across the surface of our planet. The intricacies of eclipses are as diverse as the cosmos, offering an array of phenomena, each distinct in its own right.

The variations stem from the ever-shifting position and distance of the Moon, yielding a captivating tapestry of total, partial, annular, or hybrid eclipses. Each **variation** carries its **unique** characteristics, appearances, and temporal frequencies.

Delving into this chapter, you will gain a profound comprehension of the science underpinning eclipses. You will traverse the captivating mechanics that birth these phenomena, dissecting how they transpire and uncovering the compelling art of prediction.

Through this scientific exploration, you will comprehend the profound influence eclipses wield over our planet and the celestial bodies that encircle it. As you turn the pages, you will unearth the mechanics and the captivating rhythms and patterns that govern the world of eclipses.

Eclipses extend their influence beyond the cosmic ballet in the sky. This chapter will unravel eclipses' diverse effects on the Earth, the Sun, and the Moon.

With each revelation, you'll amass a treasure trove of fascinating trivia, including why eclipses remain exclusive spectacles rather than global events, the frequencies at which these cosmic masterpieces unfold, and the historical and scientific episodes interwoven with the tapestry of eclipses.

At the end of Chapter 2, you will possess a profound comprehension of eclipses, armed with the knowledge to embark on their observation and photography. Your journey through the cosmos will continue seamlessly in Chapter 2, where the captivating history and culture of eclipses await.

Occurrence

Solar and lunar eclipses, celestial displays of mesmerizing wonder, transpire when a cosmic ballet unfolds, uniting the Sun, the Moon, and our terrestrial abode in a near-perfect alignment.

<u>Solar and Lunar Eclipses</u>

Depending on the ever-shifting position and distance of the Moon, an enchanting spectrum of eclipse varieties graces the celestial stage. Among these celestial thespians are the total, partial, annular, and hybrid eclipses, each with distinct attributes, visual allure, and temporal rhythms.

Solar Eclipses:

★ **Total Solar Eclipse:** The total eclipse reigns supreme in the realm of solar eclipses. During this extraordinary event, the Moon positions itself precisely between the Earth and the Sun, veiling the Sun's radiant face entirely.

A profound blackout ensues, casting daylight into momentary darkness. As a unique spectacle, the Sun's corona, its outer atmosphere, is unveiled, creating a faint halo encircling the Moon. Total solar eclipses are a rarity, visible only along a slender strip on Earth's surface known as the path of totality.

★ **Partial Solar Eclipse:** An alternative solar spectacle is the partial eclipse. In this celestial theater, the Moon only partially conceals the Sun, casting a partial shadow on Earth's canvas.

Its viewing domain extends beyond the narrow boundaries of totality, expanding the audience for this cosmic spectacle.

★ **Annular Solar Eclipse:** The annular solar eclipse, a beguiling spectacle, manifests when the Moon, residing at its farthest orbit from Earth, boasts an apparent size smaller than that of the Sun. In this celestial ballet, the Moon, in perfect alignment with the Sun, appears unable to obscure its entirety.

Instead, it engenders a mesmerizing sight as a slender ring of sunlight envelops the dark disk of the Moon. This celestial reverie, aptly named a "ring of fire," enraptured all who witness its fiery splendor.

★ **Hybrid Solar Eclipse:** As celestial choreography takes its course, a hybrid solar eclipse unfolds when the narrative of the eclipse metamorphoses along its path. It commences as an annular eclipse in specific locales and culminates as a total eclipse in others, or vice versa.

This fascinating celestial spectacle unfolds due to Earth's curvature and the Moon's fluctuating distances, which lead to varied degrees of coverage along the eclipse's trajectory.

Lunar Eclipses:

Lunar Eclipse

★ **Total Lunar Eclipse:** The saga of lunar eclipses commences with the total lunar eclipse, a breathtaking episode wherein the Moon enters Earth's umbral, or inner, shadow entirely. During this cosmic ballet, the Moon adopts a reddish hue, aptly dubbed a "blood moon," as Earth's atmosphere refracts sunlight, bestowing the Moon with a celestial crimson garb. The spectacle unfolds in awe-inspiring magnificence.

★ **Partial Lunar Eclipse:** The partial eclipse adorns the stage in the realm of lunar eclipses. Here, the Moon enters Earth's umbral shadow only partially, resulting in a play of shadow and light on its surface.

As some regions bask in the moonlight, others witness a transformation of the

Moon's countenance into a dusky or crimson visage.

★ **Penumbral Lunar Eclipse:** The lunar narrative culminates with the penumbral lunar eclipse, a subtle affair in which the Moon drifts into Earth's penumbral, or faint, outer shadow. A gentle dimming may be discerned during this delicate dance as the line between brightness and shadow blurs.

These celestial wonders, born of impeccable alignment and cosmic choreography, render the heavens a stage for the extraordinary. As you gaze upon these celestial performances, you venture further into the intricate tapestry of our cosmos. The narratives of solar and lunar eclipses, as described in the annals of astronomy, are merely the beginning of your celestial voyage.

Cycles and Patterns

Eclipses, those mesmerizing cosmic alignments, unfold in cycles and patterns that unveil the secrets of their predictable occurrences. These intricate rhythms, guided by celestial orchestration, allow us to anticipate when and where these spectacles will grace our skies.

<u>Eclipse Cycle and Pattern</u>

The Saros Cycle:

A star in eclipse forecasting, the Saros cycle, presents itself as a celestial timepiece, spanning approximately 18 years and 11 1/3 days. This remarkable cycle materializes due to the harmonious alignment of the Sun, the Moon, and our terrestrial abode, intricately choreographed with the Moon's orbit.

The Saros cycle comprises a sequence of eclipses bearing striking resemblances, encompassing attributes such as type, duration, and location. For instance, the annular eclipse 2023 finds its place in Saros series 134, an illustrious lineage that commenced in the annals of history in 1559 and will gracefully conclude its performance in 3049.

The Inex Cycle:

Equally noteworthy is the Inex cycle, a celestial clockwork that unfolds over approximately 29 years. This intriguing cycle emerges from the graceful shift of lunar nodes, which mark the precise points where the Moon's orbit intersects with the celestial ecliptic.

As the pages of celestial time turn, these lunar nodes trace a graceful backward path along the ecliptic, shifting by approximately 19° annually. Such celestial wanderings culminate in a poetic return to their initial positions after 29 years have unfurled.

The Inex cycle unfolds as a succession of eclipses, all sharing the same node but manifesting diverse characteristics, encompassing variations in type and location.

For instance, the total eclipse of 2024 finds its place in the rich lineage of Inex series 21, which commenced its journey in the celestial narrative of 1901 and is destined to draw its curtain in the year 2237.

These celestial timekeepers, etched in the annals of astronomy, bestow upon us the ability to anticipate and cherish the arrival of eclipses. The secrets of their cycles and patterns are woven into the cosmos, inviting us to gaze heavenward in wonder.

Effects

Eclipses, those captivating celestial phenomena, cast their spell on the observers and the elements that make up the celestial stage—the Earth, the Sun, and the Moon. These effects extend beyond mere shadows and lights, influencing the universe's physical and perceptual aspects.

Atmospheric Impact:

Eclipses profoundly impact Earth's atmosphere, creating a nuanced interplay of forces. When the Moon interposes itself between the Sun and our planet during a solar eclipse, it initiates a celestial dance with far-reaching consequences.

As the Moon's shadow sweeps over the Earth, it ushers in a temporary reduction in the Sun's ultraviolet radiation, a cosmic dimming that prompts a decline in ozone production—a gas responsible for shielding us from harmful UV rays.

Simultaneously, the eclipse imparts a touch of coolness to the shadowed realm, creating a palpable contrast in temperature between the eclipsed and non-clipsed regions. This subtle dance of temperature variation affects wind speed, humidity, and atmospheric pressure.

Tidal Symphony:

Eclipses orchestrate a celestial ballet that reverberates through the world of tides, influenced by the gravitational embrace of the Sun and the Moon.

When the Earth, Moon, and Sun align during a lunar eclipse, their gravitational interplay intensifies, resulting in tides that surge higher or retreat lower, guided by the lunar phases. In its own majestic act, the solar eclipse invokes a potent gravitational force due to the Moon's proximity to Earth.

This cosmic waltz ushers in tides that exceed their ordinary bounds dictated by the eclipse's geographical context.

Revealing Solar Secrets:

Eclipses, particularly total solar eclipses, serve as celestial veils, unveiling the Sun's hidden visage and enigmatic activity. As the Moon, a cosmic artist, completely conceals the Sun's disk, a rare opportunity emerges to witness the Sun's corona.

This tenuous but fiery outer atmosphere extends for millions of kilometers. The corona, sculpted by the Sun's magnetic embrace, graces the sky with intricate loops and streamers. This outer atmosphere also emits X-rays and ultraviolet radiation that hold sway over Earth's atmospheric milieu and its fleet of satellites.

During annular solar eclipses, where the Moon imparts a fiery ring around its disk, we are privileged to witness the Sun's surface features, from the dark sunspots, cooler regions shaped by intense magnetic activity, to the luminous faculae, hotter areas ignited by magnetic forces that stimulate heat convection.

Lunar Metamorphosis:

The Moon, a perennial companion in Earth's cosmic dance, undergoes a metamorphosis during lunar eclipses, influenced by the Sun's hidden rays. As Earth casts its shadow upon the lunar canvas, the Moon's complexion evolves into shades of gray and red, a celestial artist's palette at play.

The gray hues emanate from Earth's penumbra. This faint outer shadow dimly graces the lunar surface but refrains from complete obfuscation. The deeper shades of red are conjured by Earth's umbra, a dark inner shadow whose embrace obscures most of the Sun's rays while permitting a select few to traverse Earth's atmosphere.

This atmospheric journey scatters shorter wavelengths of light, ushering away blues and greens while bending and delivering the longer wavelengths of reds and oranges to the Moon.

Eclipses, beyond their mesmerizing visual splendor, inscribe their celestial poetry upon the annals of Earth, the Sun, and the Moon—a narrative of cosmic connections that resonate through the tapestry of the universe.

<u>Effects of Eclipses on the Earth Sun and the Moon</u>

Chapter 3

Unveiling the 2023 Annular Eclipse

Step into the realm of one of the most extraordinary celestial spectacles of the decade, the 2023 annular eclipse. It promises to etch its brilliance across the canvas of the skies, creating an unforgettable memory for all who witness it.

This remarkable event holds a special place in the hearts of skywatchers, as it not only marks a profound astronomical occurrence but also a magnificent return to the continental United States, the first since 1994. Moreover, it extends its allure to North America, offering its celestial splendor for the first time since 2012.

What sets this annular eclipse apart is its extraordinary duration. As the Moon gracefully dances between the Sun and Earth, the sky transforms into a mesmerizing display lasting for more than six minutes at its zenith, making it the longest annular eclipse of the 21st century.

This exceptional event will grace a narrow path, weaving across the United States, commencing its journey in Oregon and concluding its continental vacation in Texas. Subsequently, the celestial pageant continues over the territories of Mexico, Central America, Colombia, and Brazil, unfurling its cosmic tapestry for all to behold.

Within the chapters of this comprehensive guide, you will embark on a journey to unravel every facet of the 2023 annular eclipse. Discover the precise timings and locations to grant you an unobstructed view of this celestial waltz.

Delve into the details of the prime viewing spots, ensuring that your experience is nothing short of breathtaking. Familiarize yourself with the scientific endeavors and observations orchestrated by NASA, fostering a deeper understanding of the astronomical marvel before you. We will also introduce essential safety tips and precautions to safeguard your well-being during this awe-inspiring event.

By the end of this chapter, your readiness and excitement to witness the "ring of fire" on October 14, 2023, will be unwavering. The 2023 annular eclipse is not merely an event but an experience that transcends the boundaries of time and space. Prepare to be entranced by the heavens and to hold forever a profound appreciation for the mysteries of the cosmos.

Timings and Locations

Mark Your Calendars: The 2023 Annular Eclipse

An eagerly anticipated celestial event, the annular eclipse 2023, is set to grace our skies on Saturday, October 14, 2023. This spectacular phenomenon will cast its celestial shadow across the American continents, bestowing its radiant presence upon North, Central, and South America. The experience, however, will vary based on your geographical location and corresponding time zone.

The eclipse unveils itself in a magnificent traverse, commencing its celestial dance at one location and concluding at another. To provide a clear perspective of the eclipse timeline, we've provided the actual times in UTC when these celestial events occur.

<u>Timings and Locations of this Celestial Waltz</u>

The inaugural moment, where the partial eclipse begins, will manifest at 15:03:50 UTC, creating the first brushstroke of this cosmic masterpiece.

The final segment, the fading of the partial eclipse, will conclude at 20:55:16 UTC. Should you wish to tailor these universal times to your specific locality, an accessible Time Zone Converter is at your service, ensuring you don't miss a moment of this astronomical wonder.

The profound experience of a total eclipse will commence in Oregon at 16:10:11 UTC, marking a significant juncture in this celestial ballet. The grand finale of the total eclipse will transpire in Brazil at 19:49:01 UTC, completing its journey across the Americas.

At the peak of this extraordinary event, 17:59:32 UTC will herald the maximum point of the eclipse, an awe-inspiring moment when the Moon gracefully envelopes the most substantial portion of the Sun's radiant disk.

The path of the annular eclipse, akin to a celestial brushstroke, will traverse the United States, bestowing its grandeur upon states ranging from Oregon to Texas. Its celestial journey will continue, gracing portions of Mexico, Central America, and Colombia, concluding its continental odyssey in the vast terrain of Brazil.

As we eagerly anticipate the arrival of this celestial spectacle, mark your calendars and prepare for an experience that promises to be nothing short of mesmerizing.

The 2023 annular eclipse is a celestial gift to be cherished and observed, a reminder of the wonders of our universe. So, ready your equipment, set your alarms, and be prepared to be captivated by this celestial masterpiece on October 14, 2023.

Viewing Spots

When it comes to witnessing the annular eclipse in all its glory, location is critical. The path of annularity, where the Moon will delicately veil the Sun's disk, giving rise to the captivating "ring of fire," presents the most captivating view.

This path of annularity will chart a course across the continental United States, stretching from the picturesque landscapes of Oregon to the vibrant heart of Texas. Beyond the U.S. borders, this celestial spectacle continues, gracing portions of Mexico, Central America, Colombia, and Brazil.

Viewing Spots

It's worth noting that the annular eclipse will wear different veils when observed from various vantage points. This is mainly due to the extent to which the Moon conceals the Sun's radiant disk. Let's delve into a few illustrative examples of how this celestial phenomenon will manifest from different locales:

★ In Portland, Oregon, the eclipse unfurls its radiant dance, commencing at 9:05 a.m. PDT and concluding at 10:34 a.m. PDT. The pinnacle of this astronomical ballet unfolds at 9:50 a.m. PDT, casting the Moon as it veils 93% of the Sun's disk. This creates a mesmerizing, slender crescent of light that graces the celestial stage.

★ Across in Albuquerque, New Mexico, the eclipse takes center stage. It commences at 10:11 a.m. MDT and gracefully concludes at 12:03 p.m. MDT. At 11:08 a.m. MDT, the height of the eclipse bestows upon us a captivating spectacle as the Moon fully conceals the Sun's disk.

The result is the captivating and coveted "ring of fire".

★ Journeying to San Antonio, Texas, where the spectacle unfolds, we witness the eclipse commencing at 11:24 a.m. CDT. The celestial performance takes its final bow at 1:20 p.m. CDT, with the eclipse peak adorning the skies at 12:23 p.m. CDT. During this spellbinding moment, the Moon blankets the entirety of the Sun's radiant disk, conjuring the enchanting "ring of fire".

★ The journey continues to Mexico City, Mexico, where the eclipse commences its celestial ballet at 11:28 a.m. CDT. The final curtain falls at 1:22 p.m. CDT, with the most splendid moment gracing the skies at 12:25 p.m. CDT. The Moon then delicately enshrouds 99% of the Sun's radiant disk, creating an exquisite, almost ethereal "ring of fire".

★ Traveling onward to Bogota, Colombia, the eclipse debuted at 12:33 p.m. COT. The celestial show concludes at 2:23 p.m. COT, with the highlight of the performance illuminating the skies at 1:29 p.m. COT. During this mesmerizing segment, the Moon conceals 97% of the Sun's radiant disk, resulting in a luminous, crescent-shaped spectacle.

★ Finally, the celestial journey leads us to Brasilia, Brazil, where the eclipse graces the heavens at 2:38 p.m. BRT. The captivating performance draws to a close at 4:22 p.m. BRT, with the celestial crescendo taking place at 3:31 p.m. BRT. During this pivotal moment, the Moon veils 99% of the Sun's radiant disk, creating a strikingly slender "ring of fire" that enchants the onlookers.

As the countdown to the 2023 annular eclipse unfolds, stargazers and enthusiasts alike have the privilege of choosing their vantage points to witness this celestial masterpiece. The eclipse presents an opportunity to not only marvel at the beauty of the universe but also participate in the collective experience of celestial wonder. So, mark your calendars, ready your equipment, and choose your ideal location to participate in the October 14, 2023, grand spectacle.

Scientific Experiments and Observations

Numerous scientific experiments and observations are planned during the upcoming annular eclipse, contributing to a deeper understanding of various aspects of our solar system and planet. Some of these include:

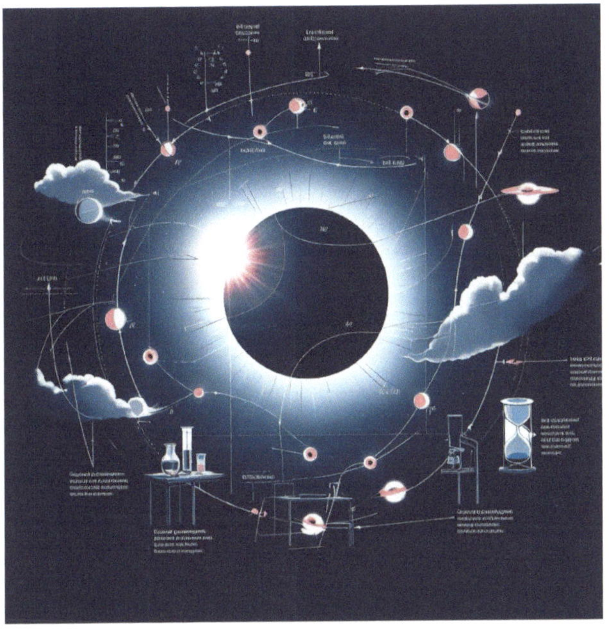

<u>Possible Scientific Event During this Event</u>

★ **Unveiling the Sun's Enigmatic Corona:**
The Sun's corona, a mysterious outer layer of its atmosphere, is typically hidden by the Sun's overwhelming brilliance but

becomes visible during an eclipse. The corona is crucial in generating the solar wind, significantly influencing Earth and other celestial bodies.

Scientists may utilize sophisticated instruments like telescopes, cameras, and spectrometers to scrutinize the corona. These instruments will help researchers measure the corona's temperature, density, magnetic field, brightness, and shape. Gaining insights into the corona will provide valuable information about the Sun's impact on our space environment and climate.

★ **Understanding Earth's Dynamic Atmosphere:** During an eclipse, the Sun's rays passing through the Earth's atmosphere create optical phenomena and illuminate the atmosphere. This unique lighting allows scientists to examine the Earth's atmosphere, understanding its composition, structure, and dynamics.

Researchers may employ diverse tools like radiosondes, high-altitude balloons, satellites, and radar systems to study atmospheric parameters such as temperature, pressure, humidity, wind patterns, and ionization. Investigating the Earth's atmosphere during an eclipse is crucial for better comprehending our planet's complex climatic processes.

★ **Observing Wildlife Behavior:** Eclipses, with their sudden alterations in lighting and temperature, can significantly impact wildlife behavior.

Animals may respond as if night or twilight has arrived, leading to changes in their activities, communication, and orientation. The study of wildlife behavior during an eclipse provides unique insights into how various species adapt to environmental transformations. Scientists will employ various techniques, including audio recordings, video cameras, and collaborative citizen science projects, to meticulously document and analyze wildlife behavior before, during, and after the eclipse. Understanding these behavioral shifts can provide vital information on how diverse species cope with environmental fluctuations.

The annular eclipse of 2023 presents a remarkable opportunity for scientific exploration, offering fresh perspectives on the Sun, Earth, and the animal kingdom. Through these experiments and observations, scientists hope to uncover new insights into the complex workings of our world and the broader cosmos.

Chapter 4

The Path Forward - Unveiling the Mysteries of Tomorrow's Eclipses

Eclipses, the celestial ballet of the Sun, Moon, and Earth, continue to captivate humanity, not just as relics of the past or captivating moments in the present but also as enigmatic events that will grace the future. In this chapter, we embark on a journey through time and space to explore what the future holds for eclipse enthusiasts and intrepid researchers.

We delve into the cosmic calendar to discover when and where the next solar and lunar eclipses will grace our skies and how, with careful planning and anticipation, we can witness these awe-inspiring events. But the future of eclipses is not just about dates and locations; it's a story of celestial choreography shaped by the intricate orbital interplay of the Earth, the Sun, and the Moon.

Interplay of the Earth, Sun and the MoonOrbital Dynamics of the earth, Sun and the Moon

As we embrace the era of space exploration, we'll also examine the boundless potential it holds for studying eclipses. From sending probes to distant celestial neighbors to the audacious dreams of human observers in alien landscapes, the future promises exciting opportunities and novel challenges.

The story of eclipses is far from over; it's a celestial narrative that continues to be written with each passing day, and we are here to witness and unveil the mysteries of tomorrow's eclipses.

When and Where

2023 promises to give us two celestial wonders - a captivating solar eclipse and an enchanting lunar eclipse. These cosmic events will grace our skies and offer opportunities for observation and contemplation. Let's delve into the details of these eclipses and how to make the most of these celestial occurrences.

1. **The Annular Solar Eclipse (October 14, 2023):**

★ An annular eclipse, a unique phenomenon, will grace our planet on October 14, 2023. This event occurs when the Moon is positioned at a distance from Earth, preventing it from completely obscuring the Sun. As a result, a glorious ring of sunlight encircles the Moon, leaving us in awe.

★ This annular eclipse will be visible from various locations, including parts of North, Central, and South America and some Pacific Ocean islands. The prime spots to witness this celestial ring in the sky include Oregon, Nevada, Utah, New Mexico, Texas, and parts of Mexico.

★ The eclipse will commence at 17:47 UTC and conclude at 21:30 UTC, reaching its pinnacle at 19:38 UTC.

To prepare for this grand spectacle:

★ Ensure your safety by acquiring special filters or devices designed for solar viewing. Gazing at the Sun without appropriate protection can cause permanent damage to your eyes. Solar glasses, pinhole projectors, telescopes with solar filters, or cameras with solar filters are safe methods for observing this eclipse.

★ Keep an eye on the weather forecast and select a location with a clear, unobstructed sky view.

2. The Partial Lunar Eclipse (October 28-29, 2023):

★ In the closing days of October 2023, the night sky will treat us to a partial lunar eclipse. During this event, the Moon will journey through Earth's shadow, revealing a striking transformation.

★ This lunar spectacle will be visible across a wide swath of the world, encompassing most of North and South America, Europe, Africa, and western Asia.

★ The eclipse will commence at 23:06 UTC on October 28, continuing until 03:28 UTC on October 29, with its zenith at 01:17 UTC.

To prepare for this lunar dance:

★ You won't require special equipment or eye protection to witness the Moon's graceful passage through Earth's shadow. To admire the lunar transformation your naked eye, binoculars, or a telescope can be employed.

★ Plan by checking the weather forecast and finding a suitable location away from the intrusion of city lights for an optimal lunar viewing experience.

In 2023, the cosmos invites you to explore these remarkable celestial events. Whether you're an astronomy enthusiast or appreciate the wonders of the universe, these eclipses offer moments of scientific insight and sheer natural beauty. Mark your calendars, gather your equipment, and set your sights on the sky for an unforgettable journey through the cosmos.

The Ever-Changing Nature of Eclipses

Eclipses, those celestial dramas that have captivated humanity for centuries, are in a state of constant transformation. Their intricate dance across the cosmic stage is choreographed by the ever-shifting interplay of the Earth, the Sun, and the Moon. Let's delve into the fascinating dynamics that make eclipses a perpetually evolving spectacle.

1. The Precession of Earth's Axis:

★ The Earth, our home in the cosmos, is not a static entity. It behaves like a spinning top, with a gentle wobble in its axis of rotation. This phenomenon, known as axial precession, carries profound implications for eclipses. Over thousands of years, the orientation of Earth's poles and equator subtly shifts. As a result, the location of eclipse paths on Earth experiences gradual yet significant changes. This cosmic waltz[1] directly influences seasonal variations and the celestial alignment during eclipses.

2. The Eccentricity of Earth's Orbit:

★ Earth's journey around the Sun is not a perfect circle; it traces an elliptical path. This orbital eccentricity leads to a fascinating interplay between the distance separating Earth from the Sun and the unfolding of solar eclipses.

★ Across the span of a year, the Earth's distance from the Sun fluctuates. Consequently, this variation directly impacts the Sun's apparent size as witnessed from our vantage point, thereby influencing the duration and spectacle of solar eclipses.

3. The Inclination of the Moon's Orbit:

★ The Moon, our constant celestial companion, engages in a celestial ballet with Earth. However, its orbit is tilted at an angle of approximately 5.1 degrees concerning the plane of Earth's orbit around the Sun, known as the ecliptic.

★ This unique orbit configuration is pivotal for eclipses, as they can only transpire when the Moon intersects the ecliptic plane. These intersections, known as nodes, are not stationary but meander along the ecliptic due to gravitational interactions with other celestial bodies.

★ This lunar orbital inclination and nodal movement significantly influence the timing and occurrence of eclipses.

4. The Eccentricity of the Moon's Orbit:

★ Just as Earth's orbit is not a perfect circle, the Moon's orbit around our planet is also elliptical. This elliptical path imparts variability to the distance between the Moon and Earth as it progresses through its monthly cycle.

★ The varying proximity between the Moon and Earth is instrumental in determining the moon's apparent size in our skies. Moreover, this proximity directly affects the type of solar eclipses we witness, whether total, annular, or hybrid.

The world of eclipses is an ever-evolving spectacle shaped by the intricate interplay of these dynamic factors. It is a captivating reminder of the dynamism and beauty inherent in our cosmic neighborhood. As we gaze upward to observe these celestial dances, we witness the celestial mechanics that have inspired awe and wonder for generations.

Challenges and Seizing

Exploring eclipses in the era of space conquest present a captivating array of challenges and prospects. As scientists and space enthusiasts embark on this cosmic journey, they grapple with the complexities of observing these rare and fleeting celestial events while also reaping the unique scientific rewards they offer.

Challenges:

1. **Precision and Planning:** Eclipses are brief and infrequent occurrences by their very nature. To capture them, meticulous planning is essential. Spacecraft and probes must be meticulously designed and launched to reach optimal locations and

orientations, ensuring that these transient phenomena are captured with precision.

<u>Spacecraft Observing an Eclipse from Space</u>

An eclipse's fleeting nature necessitates precise timing and positioning to observe and record its intricate details.

2. **Complex Celestial Interactions:** Eclipses unfold as intricate cosmic interactions between the Sun, Earth, Moon, and other celestial bodies. Spacecraft and probes must confront the harsh realities of radiation and temperature fluctuations accompanying eclipses. They must also navigate potential threats of collisions or interference from nearby celestial objects, accentuating the need for robust shielding and resilient engineering.

3. **Data Handling and Analysis:** Eclipses generate a wealth of data, necessitating effective data management. Spacecraft and probes must have dependable communication systems to transmit and receive data efficiently. Advanced storage capacity and data processing tools are

crucial for scientists and engineers to make sense of the data's complexities[3].

Opportunities:

1. **Solar Corona Studies:** Eclipses unveil a unique opportunity to scrutinize the enigmatic solar corona. This outer layer of the Sun's atmosphere is typically invisible due to the Sun's overpowering brilliance. During an eclipse, the corona becomes visible when the Moon shields the solar disk. Understanding the corona is pivotal as it is the source of the solar wind, which profoundly influences our space environment and climate. Delving into the intricacies of the corona can provide insights into how the Sun shapes our celestial neighborhood.

2. **Earth's Atmospheric Secrets:** Eclipses cast a different light on Earth's atmosphere, illuminating it in a way that is typically hidden from our view. Sunlight that traverses the Earth's atmosphere during an eclipse interacts, revealing its composition, structure, and dynamics. The atmosphere acts as a refractive lens, creating intriguing optical effects. By examining these phenomena, scientists can unravel the secrets of our planet's atmosphere, enhancing our knowledge of its intricacies and influence on our daily lives.

3. **Interplanetary Exploration:** Beyond our terrestrial home, other planets and their moons are also graced by the occurrence of eclipses. Mars, Jupiter, Saturn, and

their satellites all have celestial dances. These eclipses can offer vital insights into these distant worlds' atmospheres, surfaces, geology, and magnetism. Comparing and contrasting these findings with our home planet unveils critical information for understanding the broader cosmos.

The age of space exploration has ushered in an exciting era for studying eclipses. While these cosmic events pose challenges, they also offer rich opportunities to expand our knowledge of the Sun, Earth, and other celestial bodies that dance to the cosmic ballet of eclipses. As we continue to explore the cosmos, our understanding of these fleeting phenomena will undoubtedly deepen, adding to the tapestry of human knowledge.

Conclusion

As we reach the final pages of "2023 Annular Solar Eclipse: Ring of Fire - Black Edition," it is my hope that you have journeyed through the fascinating world of eclipses and discovered the incredible wonders they hold. We've delved into the intricate science, the rich history, the diverse cultures, and the awe-inspiring artistry that orbits these celestial events.

But this is not just the end of a book; it marks the beginning of a profound connection with the cosmos. The annular eclipse of 2023 serves as a bridge, inviting you to embark on your own personal odyssey of exploration, curiosity, and understanding.

As you prepare to witness this spectacular event, be it in the heart of the annular path or from the cozy corners of your home, remember that the universe is vast, intricate, and full of mysteries yet to be unveiled. Our world is teeming with hidden beauty, and eclipses are one of the keys to unlocking their secrets.

So, take your knowledge and passion for eclipses beyond these pages. Embrace the adventure, continue your astronomical journey, and share your newfound wisdom with the world. Seek the next eclipse, wherever it may be, and let your fascination ignite the spark of curiosity in others. Engage in scientific inquiry, artistic expression, cultural appreciation, and environmental exploration—eclipses encompass all of this and more.

With each new eclipse, our understanding deepens, our horizons expand, and our shared connection to the universe strengthens. As the shadow of the Moon dances with the light of the Sun, we, too, join in this celestial ballet, celebrating the eternal cosmic symphony.

Thank you for joining me on this enthralling eclipse expedition. May your journey be forever illuminated by the glow of knowledge, curiosity, and wonder. Here's to a future filled with eclipses that continue to inspire and astonish, connecting humanity with the stars, one celestial spectacle at a time.